YOUR KNOWLEDGE HAS VALUE

Bibliographic information published by the German National Library:

The German National Library lists this publication in the National Bibliography;
detailed bibliographic data are available on the Internet at http://dnb.dnb.de .

Imprint:

Copyright © 2015 GRIN Verlag, Open Publishing GmbH
Print and binding: Books on Demand GmbH, Norderstedt Germany
ISBN: 978-3-668-11105-9

This book at GRIN:

http://www.grin.com/en/e-book/312138/a-mathematical-approach-to-the-simple-
bulls-and-cows-code-breaking-game

Namanyay Goel, Aditya Garg

A Mathematical Approach to the Simple Bulls and Cows Code Breaking Game

GRIN Publishing

GRIN - Your knowledge has value

Since its foundation in 1998, GRIN has specialized in publishing academic texts by students, college teachers and other academics as e-book and printed book. The website www.grin.com is an ideal platform for presenting term papers, final papers, scientific essays, dissertations and specialist books.

Visit us on the internet:

http://www.grin.com/

http://www.facebook.com/grincom

http://www.twitter.com/grin_com

A Mathematical Approach to Simple Bulls and Cows

Namanyay Goel[*]   Aditya Garg[†]

DELHI PUBLIC SCHOOL VASANT KUNJ

November 26, 2015

Abstract

This document describes the game of Bulls and Cows and the research previously done on it. We then proceed to discuss our simplified algorithm which can be used practically by humans during course of play. An extended version of the algorithm, which leverages computational power to guess the code quickly and more efficiently, has also been explored. Lastly, extensive human trials have been conducted to study the effectiveness of the algorithm, and it has been shown that the algorithm results in a marked decrease in the average number of guesses in which a code is guessed by the code-breaker.

1   Introduction

The game of Bulls & Cows is played between 2 players, one of whom is the code-maker and chooses a secret code, and the other, the code-breaker, who has to guess the secret code in the minimum number of guesses.

---

[*]Electronic mail: mail@namanyayg.com
[†]Electronic mail: aditya1997@gmail.com

2 History

Bulls and Cows is a pen-and-paper code breaking game, released before and similar to the commercially popular Mastermind. The first computer program related to the game was written by Frank King at the University Of Cambridge [3] as MOO, and was similarly implemented by J. M. Grochow at MIT. The user had to enter values of Bulls & Cows, and the program used a simple predictor algorithm developed by Dr. Larmouth to guess values.

The research problem usually defined and solved involves the maximum amount of guesses it will take to guess any number. In 1976, Donald Knuth published an article showing that it takes a maximum of 5 guesses to solve a game of Mastermind [6]. B&C was extensively researched upon by Tetsuro Tanaka of University of Tokyo [4], who first found the minimum expected game length of 5.213. He also proved that the game tree for head to head play differs from that of the minimum expected game length.

3 Formalization

The code-maker writes a k-digit secret number, without repetition of digits. The code-breaker tries to guess number, and is given the number of matches after each guess. If the matching digits are in their right positions, they are "bulls", if in different positions, they are "cows". The objective of the game is to correctly guess the number in the least number of attempts.

3.1 Code

The chosen code by the code-maker is of the format $< a1, a2, a3, a4, ak >$, such that $a_i \neq a_j$ is $i \neq j$. A permutation of digits satisfying the requirement of no digits repeating is said to be a 'valid permutation' or 'possible permutation'. Hence, it can be seen that the number of possible codes can be found out using $P(10, k)$; where k is the number of digits. For $k = 3$, 720 possible codes can be created.

3.2 Guess

The codebreaker then responds with a guess (say $< b_1, b_2, b_3...b_k >$).

3.3 Response

It is given in two parts. Bulls: The code-maker has to tell the number of correct hits ($a_i = b_i$), known as bulls. Cows: The number of digits of digits present in the secret code but not at the right place ($a_i = b_j$, $i \neq j$)

There are a maximum of $\frac{(k+1)(k+2)}{2} - 1$ possible responses, which include different combinations of Bulls and Cows. We subtract 1 because if the number of bulls is k-1, the number of cows cannot be one as it has only place it can go to.

We have taken a generalization of $k = 3$, whenever needed, throughout the paper.

4 Previous Research

Optimization techniques applied to any deductive 2-player games have two objectives:

1. Minimizing the maximum numbers of guesses it can take to complete the game, or the worst case for a game.

2. Minimizing the average number of guesses it will take to guess a randomly assigned number, or the expected game length.

Known theoretical results for a generic four-place B&C game:

- The minimum expected game length is 5.213, and was proved by Tetsuro Tanaka in 1996 [4].

- The maximum number of guesses in which a number can be correctly deduced is 7, proved by multiple researchers [2].

The most optimal strategy for playing Bulls & Cows has been deduced using k-way-branching (KWB), an advanced clustering technique that heuristically obtains a strategy for the worst case scenario. Two clusters can be approximated to be equivalent if one can be derived from the other using systematic renaming of symbols. At each cluster, instead of all possible guesses, only the most possible guesses are considered, which is found using the responses given after each guess. An estimate is made for each guess, and is then used to evaluate the order in which each sub-tree is evaluated.

Class [b,c]	4,0	3,0	2,2	2,1	2,0	1,3	1,2	1,1	1,0	0,4	0,3	0,2	0,1	0,0	Total
Size	1	24	6	72	180	8	216	720	480	9	264	1260	1440	360	5040

While these techniques are highly efficient, we wanted to create our own algorithm which is can be easily explained and used practically by humans while playing the game.

5 Algorithms & Strategies

We primarily use 2 strategies, a simple strategy and then an extended strategy. The simple strategy can be explained and used by humans easily, the extended strategy requires computing power.

We use the concept of pruned set here. A pruned set is one where using on the outcomes of a guess, impossible combinations are eliminated. For example, if the code is 026 and we guess 602, we get a response of $< 0b, 3c >$, and the possibilities are pruned to $\{026, 260\}$ only.

While discussing the efficiency of any method, and making a decision regarding the next guess, the worst case scenario will always will be taken as the number of moves. *(This decision rule is known as Minimax in game theory.)* Optimum playing strategies try to eliminate the maximum number of possible values with each guess.

5.1 Responding

An initial object is created with $< 0b, 0c >$. Each digit of the guess is iterated through and is checked for presence in the code. If the digit is present, we check if it is at the right location. If it is at the right location, number of bulls is incremented otherwise number of cows is incremented.

```
function respondToNum(num, guess) {
  var response = { bulls: 0, cows: 0 };
  guess.forEach(function(dig, i) {
    dig = parseFloat(dig)
    if ( num.indexOf(dig) !== -1 ) {
      if ( num.indexOf(dig) == i )
        response.bulls++
      else
        response.cows++
```

4

```
    }
  });

  return response;
}
```

Listing 1: Responding Function

5.2 Pruning

The idea of responding can be understood as a hash function, similar to MD5/SHA-2 but less complicated, since it is a one-way function.

Hence, we use a simple brute-force approach to prune a set. We test all responses from all possible permutations, and compare it against the actual response. We add all permutations that give a matching response to the new set, which we deliver as the pruned set on completion. The code is described below

```
function pruneSet(set, guess, ans) {
  var response = respondToNum(ans, guess);
  var pruned = [];
  set.forEach(function(num, pos) {
    var numRes = respondToNum(num, guess)
    if ( numRes.bulls == response.bulls && numRes.cows == response.
    cows ) {
      pruned.push(num)
    }
  });
  return pruned;
}
```

Listing 2: Pruning Function

5.3 Human-friendly strategy

The important thing to is that it isn't always necessary that the guess, which eliminates the most number of possible values, and prunes the set the most, belongs the the currently pruned set.

For example, while guessing the number 798, a code-breaker arrives on the set $\{098, 198, 298, 398, 498, 598, 698, 798\}$. If numbers from the set are selected and guessed, a maximum of 8 attempts might be required before the final answer is required.

5

Instead, guessing 012 will lead to the response $< 0b, 0c >$ and set being pruned to $\{398, 498, 598, 698, 798\}$. Again guessing 345 will similarly reduce the set to $\{698, 798\}$. Thus the maximum number of attempts gets reduced to 4.

Statistically, the second approach is better. Formalizing the algorithm mathematically, we find that it requires the following steps:

1. Order digits by frequency as they occur in the pruned set.

2. Choose the digits that occur least frequently and form a guess with them.

This strategy is, of course, rudimentary. However, based on our human studies, we were able to find good success rates after explaining this algorithm.

We wrote a JavaScript program to simulate this strategy. Each digit is represented as an object which lists the value, position, and frequency the digit has occurred.

`getHashTable` function converts a given set into the above number representation. `searchInHash` function searches the existing records to check if the digit is represented already, and if so, increments the frequency by 1. Otherwise, a new record is created.

The records are then sorted in ascending order of digit frequency and passed to the `makeGuess` function which create a valid guess without repeats out of sorted records.

```
function getHashTable (s) {
  var hash = [];
  s.forEach(function(num) {
    for ( var i = 0; i < 3; i++ ) {
      var res = searchInHash(hash, i, num[i])
      if ( res ) res.count++
      else hash.push({ pos: i, val: num[i], count: 1 })
    }
  });

  hash.sort(function(a, b) {
    if ( a.count > b.count ) return 1
    if ( a.count < b.count ) return -1
    return 0;
  })

  return hash;
}

function makeGuess (s) {
  if ( s.length == 1 )
    return s;
```

```
    var sortedHash = getHashTable(s);
25  var num = [];

27  var i = 0;

29  while ( num.length !== N ) {
      if ( num.indexOf(sortedHash[i].val) == -1 )
31      num.push(sortedHash[i].val)
      i++;
33  }

35  return num;
}
```

Listing 3: Human-friendly Strategy Implementation

No. of Guesses	1	2	3	4	5	6	7	8	9
Frequency	1	3	8	44	156	250	190	56	12

Table 1: Frequency chart for human-friendly strategy

Frequency of guesses required by human-friendly strategy

Average guesses taken: 6.0736

5.4   Extended Strategy

What we observe is that the human strategy tries to figure out a number that is able to eliminate the maximum number of guesses, but in a crude way. Leveraging computing power, the algorithm can check through all possible permutations to find number(s), that when guessed, are able to reduce the pruned set to only one possibility.

The way it works is that loops through all possible permutations of 3 digit numbers, and assumes they are 'guesses'. It then checks what response is found if all numbers of the existing pruned set are taken as 'codes' and are guessed for using the permutation. The responses are stored and compared to each other. If a permutation is able to act as a unique guess when all the items of the pruned set are taken as codes, it is considered to be a 'unique permutation' and the value is used to make the real guess. This strategy ensures that the resulting set is pruned to only one possibility once the unique permutation is taken as a real guess.

To take an example, consider the pruned set $\{243, 342, 432, 425, 142, 153\}$. If we check the lowest frequency using the human-friendly strategy, we find that the numbers 1 and 5 occur the least number of times. The primitive approach would try to use those numbers in the guess. However, this pruned set has one guess that will prune the set to one element after the response: 243. Evaluating this guess for all possible codes from the pruned set, we see:

- If the code was 243, we'd get $< 3b, 0c >$

- If the code was 342, we'd get $< 1b, 2c >$

- If the code was 432, we'd get $< 0b, 3c >$

- If the code was 425, we'd get $< 0b, 2c >$

- If the code was 142, we'd get $< 1b, 1c >$

- If the code was 153, we'd get $< 1b, 0c >$

Since we've arrived at a unique permutation, we're able to predict the code is based on the response we get. Hence, we get an answer in 2 turns.

The code below uses `allUniques`, an array of all possible permutations. It runs `respondToNum` on all members of the set, treating them as the secret code to see the response. It then converts the response into a string and pushes it into an array.

This array is then checked for matches with a simple algorithm. Each element of the array is compared to each element of the array, and the number of matches is added to a variable. If the value of the variable equals to the length of the array, it means that array has unique elements. Here, we initialize the variable with the value of -(array length) to reach '0' in case of no matches.

If there successfully are no matches, the number is added to an array 'uniques', which is returned to the program once all numbers have been evaluated in this way.

```javascript
function findUniqueGuess (set) {
  if ( set.length == 0 ) return set[0];

  var responses = [];
  var uniques = [];

  allUniques.forEach(function(number) {
    var num = number.split('');

    set.forEach(function(n) {
      var res = respondToNum(n, num);
      var resStr = 'b' + res.bulls + 'c' + res.cows;
      responses.push(resStr);
    });


    var matches = -responses.length;
    responses.forEach(function(res) {
      responses.forEach(function(r) {
        if ( res == r ) matches++
      });
    })
    if ( matches == 0 ) {
      uniques.push(num);
    }

    responses = [];
  });
  return uniques;
}
```

Listing 4: Extended Strategy Implementation

No. of Guesses	1	2	3	4	5	6	7	8	9
Frequency	1	2	7	33	157	359	142	19	0

Table 2: Frequency chart for extended strategy

Average Guesses Required: 5.8930

This algorithm has been used in the implementation of our web application[1].

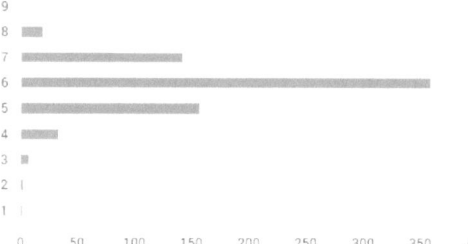

Frequency of guesses required by extended strategy

6 Human Psychology

6.1 Method

We conducted tests with k = 2 and k = 3 with human subjects of varying ages, and changed the amount of information given to them. We began with merely explaining the rules to the subjects, but after Tests C, we explained the human-friendly strategy.

Information refers to the information given to the subject after a guess. To assist showing pruned set in Tests B and D, we used the web application [1].

Note that all games was played mentally, without using any means to record information.

6.1.1 Test Series I: Subject Age 11-15 years

Tests A:
Information: Number of bulls and cows.

The subject chose random numbers, and even numbers they 'liked', to begin with. This continued until they reached 2-3 cows, after which their strategy changed to using permutations of existing numbers to figure out bulls. On reaching 2 bulls, they guessed until they found the number.

Guesses required: 20 - 24

Tests B:
Information: Number of bulls and cows; pruned set.

Seeing the pruned set resulted in quicker convergence after reaching around 3 hits. Initial strategy remained the same.

Guesses required: 18 - 21

Tests C: Strategy Explained
Information: Number of bulls and cows

After explaining the human-friendly strategy, subject was able to make better decisions when they reached a smaller set. However, the average number of attempts was similar to average number of attempts in Tests B, perhaps due to the subjects subconsciously coming up with this algorithm by themselves, or having difficulty in visualizing pruned sets.

Guesses required: 17 - 19

Tests D:
Information: Number of bulls and cows; pruned set.

Once the subjects were able to see the pruned set, they were able to come up with better guesses. However, they were not able to do that if the set size was large. Nonetheless, number of guesses required reduced considerably.

Guesses required: 10 - 12

6.1.2   Test Series II: Subject Age 35-45 years

Tests A:
Information: Number of bulls and cows.

The subjects chose random numbers initially, but were able to narrow down results better as they approached larger number of hits.

Guesses required: 15 - 18

Tests B:
Information: Number of bulls and cows; pruned set.

The pruned set allowed the subjects to reach a faster conclusion when set size was smaller. It seemed that they had come up with an algorithm mentally, albeit one that performed poorer than our algorithm.

Guesses required: 13 - 16

Tests C: Strategy Explained
Information: Number of bulls and cows

After explaining the human-friendly strategy, subject made better decisions right from the beginning. They began choosing numbers that did not necessarily occur in the pruned set, but those that sufficiently reduced the set size.

Guesses required: 9 - 12

Tests D:
Information: Number of bulls and cows; pruned set.

Seeing the pruned set allowed the subjects to come to a conclusion much faster. They took more time in evaluating the steps and thinking about the perfect number that would reduce the possibilities to the maximum amount.

Guesses required: 7 - 11

7 Conclusion

A simple derivation and explanation of a strong algorithm for the game is concluded to be possible across age groups. Programming the same strategies has proven to be viable as well.

The primitive algorithm and extended algorithm have a slight but important difference in average case scenario which cannot be ignored. However, a large amount of processing power is needed for the extended approach, due to the NP-hard nature of the problem.

A vast number of approaches and strategies for the game still remain unsolved. Nonetheless, thanks to the brevity of its rules, Bulls and Cows poses a very interesting challenge to recreational mathematicians, computer engineers, and psychologists alike.

References

[1] Implementation of described algorithms
http://projects.namanyayg.com/moo

[2] John Francis, *Strategies for playing MOO, or "Bulls and Cows"*, http://slovesnov.users.sourceforge.net/bullscows/bulls_and_cows.pdf

[3] John Wiley & Sons, Ltd., *MOO in Multics*, http://www.multicians.org/moo-in-multics-1972.pdf

[4] Tetsuro Tanaka, *An optimal MOO strategy*, http://dell.tanaka.ecc.u-tokyo.ac.jp/~ktanaka/papers/gpw96.pdf

[5] Minimax Algorithm, http://web.stanford.edu/~msirota/soco/minimax.html

[6] Donald E. Knuth *The Computer as Master Mind* http://www.dcc.fc.up.pt/~sssousa/RM09101.pdf

YOUR KNOWLEDGE HAS VALUE